2024 First Printing

Growing God

Copyright © 2024 by Karen Kiefer

ISBN 978-1-64060-908-2

The Paraclete Press name and logo (dove on cross) are trademarks of Paraclete Press, Inc.

Library of Congress Cataloging-in-Publication Data
Names: Kiefer, Karen, 1960- author. | Wit, Kathy de, illustrator.
Title: Growing God / by Karen Kiefer ; illustrated by Kathy De Wit.
Description: Brewster, Massachusetts : Paraclete Press, [2024] | Summary: "Emma's Nana gave her seeds and told her to plant them and be patient, as she was Growing God. Through her perseverance, even her skeptical friends believed her, as they saw the seeds grow into living plants, a gift from God"-- Provided by publisher.
Identifiers: LCCN 2023030665 (print) | LCCN 2023030666 (ebook) | ISBN 9781640609082 (hardcover) | ISBN 9781640609099 (epub) | ISBN 9781640609105 (pdf)
Subjects: LCSH: Growth (Plants)--Religious aspects--Christianity--Juvenile literature. | God (Christianity)--Juvenile literature.
Classification: LCC QK49 .K49 2024 (print) | LCC QK49 (ebook) | DDC 581--dc23/eng/20230907
LC record available at https://lccn.loc.gov/2023030665
LC ebook record available at https://lccn.loc.gov/2023030666

10 9 8 7 6 5 4 3 2 1
All rights reserved. No portion of this book may be reproduced, stored in an electronic retrieval system, or transmitted in any form or by any means—electronic, mechanical, photocopy, recording, or any other—except for brief quotations in printed reviews, without the prior permission of the publisher.

Manufactured by Shenzhen Tianhong Printing Co. Ltd.
Printed October 2023, in Pinghu, Shenzhen, China
This product conforms to all applicable CPSIA standards.
Batch: 202310002STH

Published by Paraclete Press
Brewster, Massachusetts
www.paracletepress.com

*To the Emma that lives within each one of us,
may we continue to be a seed for others
and find new opportunities to grow in God's love.*

KAREN

*For our two most beautiful seeds planted
in 2014, who keep flowering our lives, and for my
husband who keeps me flowering, always.*

KATHY

I stayed at my favorite place on earth during school vacation: my Nana's farm. Every day, Nana and I went out to the fields to water, plant, and pluck flowers and vegetables.

Nana told me the farm is where she grows God and that every seed holds an opportunity. I wasn't really sure what she meant, but I knew if Nana's farm could make my heart feel this happy, I wanted to grow God, too.

I think Nana had a hunch, because a few days after I got home, she sent me a humongous box filled with lots of seeds and a note:

My dearest Emma,

God lives in these seeds. Plant them, love them, be patient, and believe. Wait until you see what happens when you care for God's creation.

Love and hugs, Nana

Looking at the itty-bitty seeds, I whispered, "Hello, God."

The next morning, I found the perfect spot in the yard to plant the seeds. Staring at the dirt, I imagined blossoms of peonies, pansies, sunflowers, and vegetables galore.

I dug deep and turned the soil just like Nana. I tucked the seeds in just so, making sure the dirt could hug them. When I finished planting the last seed, I said, "Don't forget to dance with the sun and sleep with the moon!"

Then, I heard nagging, know-it-all Nigel yell, "Emma, are you talking to the dirt?" "Yes," I said proudly, "I'm growing GOD." Nigel looked at me in disbelief. "You are what?" he said, laughing. "That's right, Nigel, just wait." As Nigel walked away, he shouted, "Oh, Emma, believe me, I'll wait."

Days passed, and I knew in my heart the seeds were calling me to care for them. I watered them, talked to them, prayed over them, and reminded them, "Can't wait to meet you."

The sun and the moon helped me watch over the seeds, and the rain visited on occasion. At night I prayed, "God, please help me grow you." And during the day I tried to practice patience, but it had already been 33 days, and the dirt was still as black as night. No sign of green, not even a tiny sprout.

Meanwhile, Nigel let every kid in the neighborhood know that I was trying to grow God. They would stop by to stare and snicker, and Nigel would chime in, "Emma, I'm waiting."

To be honest, I was waiting too. I wondered, "Is something wrong? Maybe the dirt was hugging the seeds too tightly or maybe the seeds were too shy to poke through and say hello?" Then I reread Nana's note, reminding me to be patient and believe.

Then, one glorious morning, it happened. A green sprout pushed its way through the dirt. "Hello, sweet sprout," I cheered. It stretched out like it was trying to hug me. "I see you, God," I cried, in awe of this beautiful creation.

I didn't know Nigel was standing behind me until he growled, "Emma, that's not God, it's a puny sprout." I yelled back, "God lives in this sprout," but Nigel was already blocking his ears.

As soon as the sun woke, I ran to my sprout and my heart fell to the ground. It had already turned brown and toppled over. I wanted to wail, "NO, sweet one, you can't be dead!" But instead, I held my cry in with all my might. I knew I couldn't let Nigel or anyone else hear me.

Panicked, I quickly called Nana, but could barely talk through my tears. "Emma," she said, "get some cornmeal from the kitchen and sprinkle it over the dirt. It will absorb excess moisture and stop your other spouts from dying."

Before she hung up, Nana reminded me, "Emma, every seed holds an opportunity. Even the seeds that don't live long have something to teach us."

I sprinkled the dirt with cornmeal, but not before every kid in the neighborhood found out about my dead sprout. "Emma, are you sprinkling fairy dust to make God grow?" Nigel yelped. The neighborhood kids were laughing and pointing.

I should have been mad, but I wasn't. I should have hollered, "God doesn't need fairy dust," but I didn't. I don't know how to explain it, but it was as if God planted a seed deep inside of me and it was taking root. I knew I was growing God.

More sprout-less days followed. Then, one glorious morning, there were sprouts everywhere, and I mean everywhere, even in places where I didn't plant seeds. "I see you, God," I cheered.